SALINA FREE LIBRARY
100 BELMONT STREET
MATTYDALE, NY 13211

TOILETS
from Then to Now

BY RACHEL GRACK

AMICUS | AMICUS INK

Sequence is published by Amicus and Amicus Ink
P.O. Box 1329, Mankato, MN 56002
www.amicuspublishing.us

Copyright © 2020 Amicus. International copyright reserved in all countries. No part of this book may be reproduced in any form without written permission from the publisher.

Library of Congress Cataloging-in-Publication Data
Names: Koestler-Grack, Rachel A., 1973- author.
Title: Toilets from then to now / by Rachel Grack.
Description: Mankato, MN : Amicus/Amicus Ink, [2020] | Series: Sequence developments in technology | Includes bibliographical references and index. | Audience: Grade 4-6.
Identifiers: LCCN 2018048933 (print) | LCCN 2018049523 (ebook) | ISBN 9781681517681 (pdf) | ISBN 9781681516868 (library binding) | ISBN 9781681524726 (pbk.)
Subjects: LCSH: Toilets--History--Juvenile literature.
Classification: LCC TH6498 (ebook) | LCC TH6498 .K67 2020 (print) | DDC 696/.182--dc23
LC record available at https://lccn.loc.gov/2018048933

Editor: Wendy Dieker
Designer: Aubrey Harper
Photo Researcher: Holly Young

Photo Credits: Rasulovs/iStock cover; KOHb/iStock cover; Zeiss4Me/iStock 4; BMPix/iStock 6–7; Archiv Gerstenberg/ullstein bild/Getty 8; Roi.dagobert/WikiCommons 10–11; unknown photographer/Wordpress/The Street and the City/Awakenings 12; World History Archive/Alamy 14–15; Retrograph Collection/Mary Evans 16; The Printing Art 1913 Volume 20/Mary Evans 18–19; grandriver/Getty 20–21; CUHRIG/iStock 23; Yoshikazu Tsuno/AFP/Getty 24; Bloomberg/Getty 26–27; Mahesh Kumar A/AP 28–29

Printed in the United States of America

HC 10 9 8 7 6 5 4 3 2 1
PB 10 9 8 7 6 5 4 3 2 1

TABLE OF CONTENTS

We Need Toilets	5
Early Flush Toilets	13
Water Closets and Outhouses	17
Toilet Technology	22
Toilets for All	29

■ ■ ■ ■ ■

Glossary	30
Read More	31
Websites	31
Index	32

Around the world, people know they will find toilets when they see a sign like this.

We Need Toilets

People don't think about toilets until they need to use one. Most buildings in developed countries have indoor bathrooms with flushing toilets. Imagine living in a place with no running water or **sanitation** systems. Without these things, human waste is messy and smelly. It can spread deadly diseases. Throughout history, people have worked to find ways to make "going" better.

The idea of flushing toilets is not new. Around 1700 **BCE**, Greeks built a palace with the first flushing toilets. They were connected to a **sewer** system of stone pipes. To flush, people dumped water down the hole. The water carried waste away through the pipes. Few places copied the Greeks, though.

> This manually flushing toilet is from hundreds of years after the Greeks built theirs.

A Greek palace features the world's first flushing toilets.

1700 BCE

. . LOADING . . LOADING .

Some people used to empty their waste into the streets.

A Greek palace features the world's first flushing toilets.

1700 BCE

Around 500 CE

People empty chamber pots into town cesspits.

Elsewhere, most people used **chamber pots**. In the Middle Ages (500–1500 CE), workers called nightmen collected the waste from homes. They dumped it into a **cesspit**. These pools of waste were stinky and dangerous. People sometimes fell in and drowned. Some didn't wait for the nightmen. They just emptied their pots in the street. It was not a clean time.

By 1100, castles in Europe had **garderobes**. These bathrooms looked like closets hanging out of the walls. Inside was a bench with a hole. The bottomless toilet dropped poop to the ground. Some garderobes emptied into an underground cesspit. Some emptied into the **moat** around the castle. No wonder moats turned enemies away! Stinky!

> A garderobe still hangs off the side of a castle in France.

1700 BCE	Around 500 CE	1100
A Greek palace features the world's first flushing toilets.	People empty chamber pots into town cesspits.	Castles feature bathrooms called garderobes.

G...LOADING.

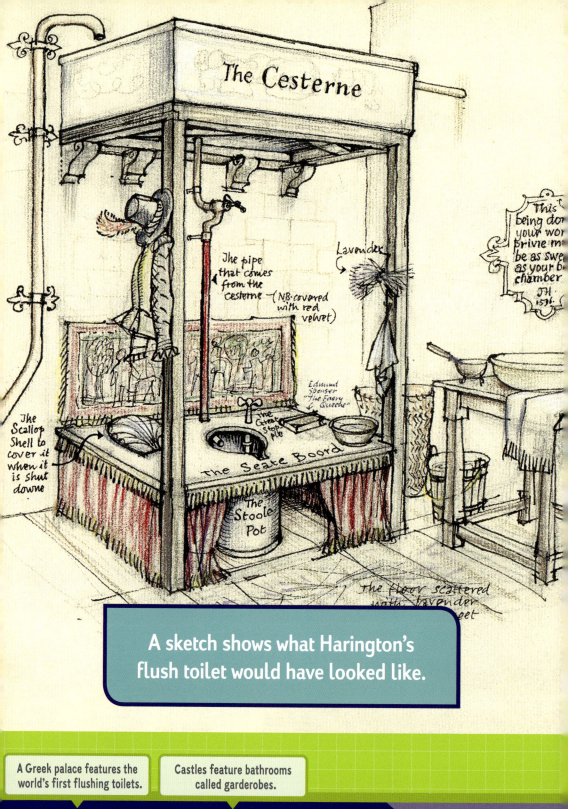

A sketch shows what Harington's flush toilet would have looked like.

A Greek palace features the world's first flushing toilets.	Castles feature bathrooms called garderobes.		
1700 BCE	Around 500 CE	1100	1596
	People empty chamber pots into town cesspits.		Sir John Harington invents the modern flush toilet.

Early Flush Toilets

Englishman Sir John Harington invented the first modern flush toilet in 1596. It had a tank that sent water to flush out the bowl. The waste went out a drain. He built one for himself and another for the Queen of England. But people didn't have running water at that time. His toilet didn't catch on.

By the late 1700s, cities in Europe started putting in plumbing systems. These pipes brought water into buildings. This made it easier to fill a toilet tank. Inventor Alexander Cumming got to work. In 1775, he came up with a flushing toilet design. It had a bowl filled with water and an S-shaped drain.

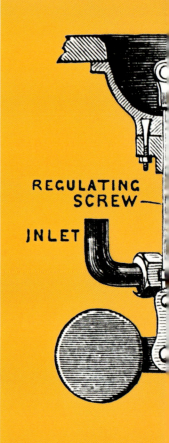

A diagram from the 1800s shows Cumming's basic design with small improvements.

A Greek palace features the world's first flushing toilets.		Castles feature bathrooms called garderobes.		Alexander Cumming invents the flush toilet used today.	
1700 BCE	Around 500 CE	1100	1596	1775	
	People empty chamber pots into town cesspits.		Sir John Harington invents the modern flush toilet.		

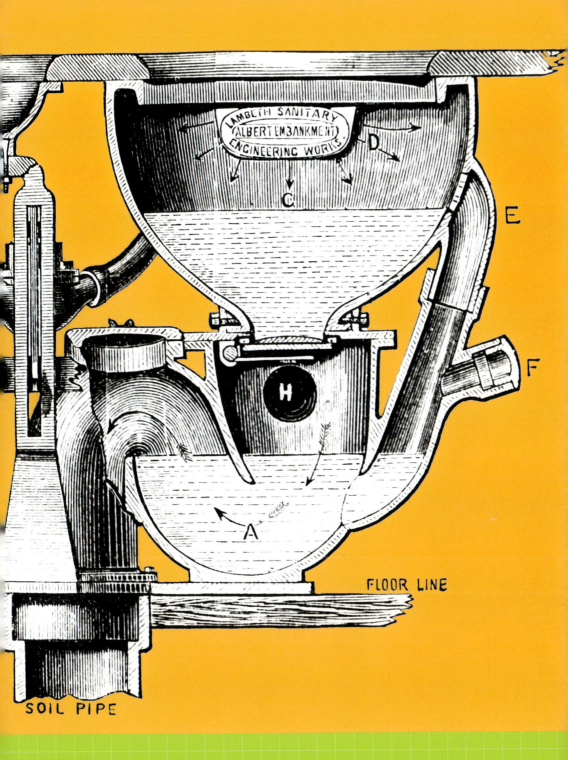

People had to pull a chain to let water flow from the high tank to flush the bowl.

A Greek palace features the world's first flushing toilets.		Castles feature bathrooms called garderobes.		Alexander Cumming invents the flush toilet used today.	
1700 BCE	Around 500 CE	1100	1596	1775	1880s
	People empty chamber pots into town cesspits.		Sir John Harington invents the modern flush toilet.		North American homes feature high-tank toilets.

16

Water Closets and Outhouses

Plumbing still cost a lot of money to put in. It took until the 1880s for cities in North America to begin building water systems. Then more people were putting flush toilets in their homes. These toilets sat in a little room called a **water closet**. A tank was mounted high on the wall. It used gravity to help the flush work.

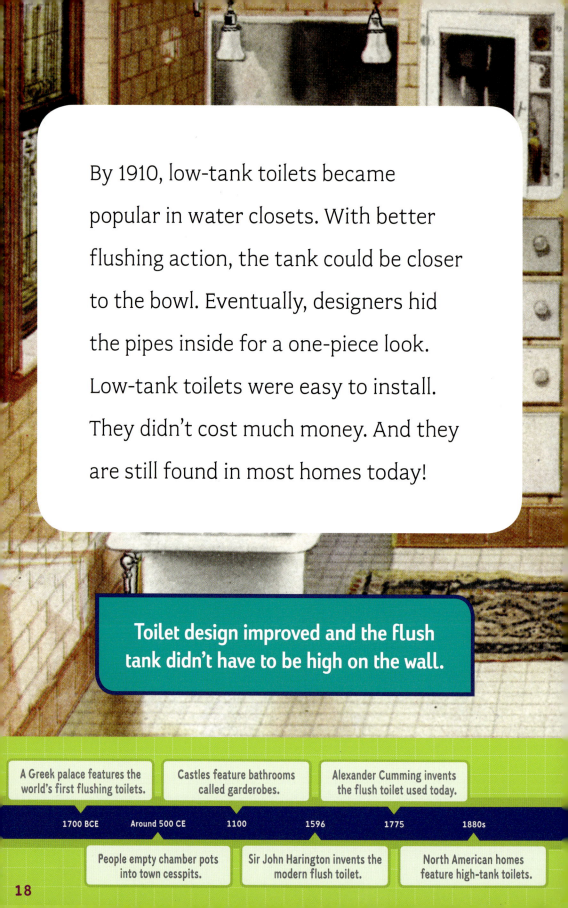

By 1910, low-tank toilets became popular in water closets. With better flushing action, the tank could be closer to the bowl. Eventually, designers hid the pipes inside for a one-piece look. Low-tank toilets were easy to install. They didn't cost much money. And they are still found in most homes today!

Toilet design improved and the flush tank didn't have to be high on the wall.

A Greek palace features the world's first flushing toilets.		Castles feature bathrooms called garderobes.			Alexander Cumming invents the flush toilet used today.	
1700 BCE	Around 500 CE	1100	1596		1775	1880s
	People empty chamber pots into town cesspits.		Sir John Harington invents the modern flush toilet.			North American homes feature high-tank toilets.

1910
Low-tank toilets become the most popular toilet style.

- A Greek palace features the world's first flushing toilets. — 1700 BCE
- People empty chamber pots into town cesspits. — Around 500 CE
- Castles feature bathrooms called garderobes. — 1100
- Sir John Harington invents the modern flush toilet. — 1596
- Alexander Cumming invents the flush toilet used today. — 1775
- North American homes feature high-tank toilets. — 1880s

It took a long time for indoor plumbing to reach **rural** areas. People were still using outhouses. In the 1930s, the Work Projects Administration (WPA) took on a big project. They updated over 2 million outhouses. The new outhouses had concrete waste pits. They were used into the 1960s. By then, indoor plumbing was in almost all U.S. homes.

> Before people had running water, they used toilets in outhouses.

The WPA builds over 2 million outhouses across America.

1910   1930s

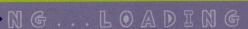

Low-tank toilets become the most popular toilet style.

21

Toilet Technology

By 1990, many U.S. homes had more than one bathroom. That's a lot of water getting flushed down the toilet! Back then, toilets used up to 5 gallons (19 L) of water each flush. The Energy Policy Act of 1992 helped save water. It says that new toilets need to be low-flow. These toilets use 1.6 gallons (6 L) of water or less.

Flushed water goes through a treatment plant to be safely dumped or reused.

1700 BCE	Around 500 CE	1100	1596	1775	1880s
A Greek palace features the world's first flushing toilets.		Castles feature bathrooms called garderobes.		Alexander Cumming invents the flush toilet used today.	
	People empty chamber pots into town cesspits.		Sir John Harington invents the modern flush toilet.		North American homes feature high-tank toilets.

The WPA builds over 2 million outhouses across America.

1910 — Low-tank toilets become the most popular toilet style.

1930s

1992 — The Energy Policy Act requires new toilets to be made low-flow.

...LOADING...

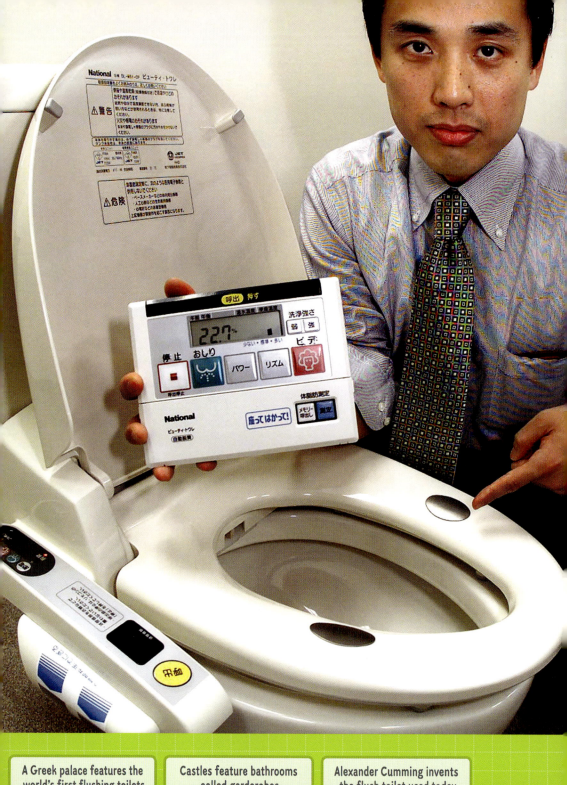

A Greek palace features the world's first flushing toilets.		Castles feature bathrooms called garderobes.		Alexander Cumming invents the flush toilet used today.	
1700 BCE	Around 500 CE	1100	1596	1775	1880s
	People empty chamber pots into town cesspits.		Sir John Harington invents the modern flush toilet.		North American homes feature high-tank toilets.

Today, toilets do more than save water. They can make "going" more enjoyable. Since 2002, Matsushita toilets have the latest in luxury and comforts. Stools feature lids that open on their own. They have **automatic** flushing. Some have automatic cleaning cycles. High-end toilets light up and have air fresheners. The seats can even heat up.

A remote control lets toilet users pick comfortable settings.

| 1910 | 1930s | 1992 | 2002 |

- The WPA builds over 2 million outhouses across America.
- Matsushita creates a luxury line of toilets.
- Low-tank toilets become the most popular toilet style.
- The Energy Policy Act requires new toilets to be made low-flow.

Toilets continue to get more comfortable—and smarter. In 2018, Kohler created the Numi intelligent toilet. Its seat warms, washes, and dries your bottom. A remote saves your settings for the next visit. You can talk to this toilet, too! It can be controlled by your voice. It even has speakers that can play your favorite tunes.

> The Numi toilet is displayed at an electronics show. Who ever thought toilets could be so high tech?

1700 BCE	Around 500 CE	1100	1596	1775	1880s
A Greek palace features the world's first flushing toilets.		Castles feature bathrooms called garderobes.		Alexander Cumming invents the flush toilet used today.	
	People empty chamber pots into town cesspits.		Sir John Harington invents the modern flush toilet.		North American homes feature high-tank toilets.

	The WPA builds over 2 million outhouses across America.		Matsushita creates a luxury line of toilets.		
1910	1930s	1992	2002	2018	
Low-tank toilets become the most popular toilet style.		The Energy Policy Act requires new toilets to be made low-flow.		Kohler offers an intelligent toilet with voice-command.	

I AM ONE IN 2.4 billion People in the world having No Toilet

A Greek palace features the world's first flushing toilets.		Castles feature bathrooms called garderobes.		Alexander Cumming invents the flush toilet used today.		
1700 BCE	Around 500 CE	1100	1596	1775	1880s	
	People empty chamber pots into town cesspits.		Sir John Harington invents the modern flush toilet.		North American homes feature high-tank toilets.	

Toilets for All

Some people enjoy heated toilet seats. But about one-third of the people in the world do not have proper toilets. In 2013, the **United Nations** created World Toilet Day. It is a way to tell people about the sanitation needs of others around the world. The U.N. keeps working to build clean bathrooms for everyone.

Girls rally in India to tell people that they do not have toilets.

1910	1930s	1992	2002	2018
Low-tank toilets become the most popular toilet style.	The WPA builds over 2 million outhouses across America.	The Energy Policy Act requires new toilets to be made low-flow.	Matsushita creates a luxury line of toilets.	Kohler offers an intelligent toilet with voice-command.

Glossary

automatic A machine that works on its own without someone controlling it.

BCE Short for Before the Common Era; it notates years before year 0 using the Gregorian calendar; years after 0 are called the Common Era (CE).

cesspit A pit used for dumping human waste.

chamber pots Pots or bowls used as indoor toilets to be dumped out later.

garderobes A closet-like room hanging off the outside wall of a castle; waste was dropped to the ground below.

moat A ditch or canal dug around a castle to work as an obstacle for enemies.

rural Having to do with a place that is outside of a town or city.

sanitation To do with cleanliness and the conditions of clean water and waste removal.

sewer An underground pipe that carries away waste from sinks, baths, and toilets.

United Nations An organization that works to keep friendly relationships between countries in the world and help countries overcome challenges that affect their citizens.

water closet Another name for early toilets or bathrooms.

Read More

Abell, Tracy. *Toilets.* Lake Elmo, Minn.: Focus Readers, 2018.

Perdew, Laura. *How the Toilet Changed History.* Minneapolis: Abdo Publishing, 2015.

Roza, Greg. *How Do Sewers Work?* New York: PowerKids Press, 2016.

Websites

Curious Kids: Where Does My Poo Go When I Flush the Toilet?
https://theconversation.com/curious-kids-where-does-my-poo-go-when-i-flush-the-toilet-does-it-go-into-the-ocean-78254

It's Okay to Be Smart: How the Toilet Changed History
www.youtube.com/watch?v=GWQG1YZS9l4

World Toilet Day
www.worldtoiletday.info

Every effort has been made to ensure that these websites are appropriate for children. However, because of the nature of the Internet, it is impossible to guarantee that these sites will remain active indefinitely or that their contents will not be altered.

Index

castle toilets 10
cesspits 9, 10
chamber pots 9
Cumming, Alexander 14
diseases 5
Energy Policy Act 22
flushing toilets 5, 6, 13, 14, 17, 18, 22, 25
garderobes 10
Greek toilets 6
Harington, Sir John 13
high-tank toilets 17
intelligent toilets 25, 26
low-flow toilets 22
low-tank toilets 18
Middle Ages 9
nightmen 9
outhouses 21
plumbing systems 14, 17, 21
rural areas 21
saving water 22, 25
sewer systems 6
water closets 17, 18
Work Projects Administration (WPA) 21
World Toilet Day 29

About the Author

Rachel Grack has worked in children's nonfiction publishing since 1999. Rachel lives on a small desert ranch in Arizona. She enjoys spending time with her family and barnyard of animals. Thanks to our wireless world, her ranch stays tapped into developing technology.